WEST CHICAGO PUBLIC LIBRARY DISTRICT
3 6653 00236 5041

W9-BPL-278

REVOLUTIONARY DISCOVERIES OF SCIENTIFIC PIONEERS™

THE DOUBLE HELIX STRUCTURE OF DNA: JAMES WATSON, FRANCIS CRICK, MAURICE WILKINS, AND ROSALIND FRANKLIN

R. N. ALBRIGHT

Published in 2014 by The Rosen Publishing Group, Inc.
29 East 21st Street, New York, NY 10010

First Edition

Library of Congress Cataloging-in-Publication Data

Albright, R. N.
The double helix structure of DNA: James Watson, Francis Crick, Maurice Wilkins, and Rosalind Franklin/R.N. Albright.—First edition.
pages cm.—(Revolutionary discoveries of scientific pioneers)
Audience: Grades 7–12.
Includes bibliographical references and index.
ISBN 978-1-4777-1809-4 (library binding)
1. DNA—Juvenile literature. 2. DNA—Research—History—Juvenile literature.
3. Molecular biologists—Biography—Juvenile literature. I. Title.
QP624.A415 2014
572.8'6—dc23

2013023144

Manufactured in the United States of America

CPSIA Compliance Information: Batch #W14YA: For further information, contact Rosen Publishing, New York, New York, at 1-800-237-9932.
A portion of the material in this book has been derived from *Watson and Crick and DNA* by Christy Marx.

CONTENTS

INTRODUCTION

This is the story of a molecule—not just any molecule, but one that contains the code of life. That molecule is called deoxyribonucleic acid, or DNA.

DNA is found inside every living organism on Earth, and it serves as a chemical instruction book for the way an organism develops and functions. The language of that book has only four "letters," called nucleotides. The instructions in DNA tell the organism whether to have many cells or one cell. They tell it whether to be a bacterium, a fungus, a plant, or an animal (or a member of the less familiar kingdoms of life called archaea and protists). They tell it what species of living creature to be.

They even tell the organism how to be its particular self. If it is a pea plant, those instructions tell it whether to grow tall or short, have white or purple flowers, or produce smooth or wrinkled peas. If it is a dog, they tell it whether to be a male Saint Bernard or a female dachshund. If it is a human, they tell it what color eyes to have, what texture of hair to have, and what special abilities it may develop that set it apart from others.

Of course a person's special abilities do not depend on DNA alone. They also depend on the environment in which that person lives and the influences of other

THIS DIAGRAM SHOWS THE SPIRAL-STAIRCASE STRUCTURE OF A DNA MOLECULE. DISCOVERING THAT STRUCTURE AND HOW THE MOLECULE CAN SERVE AS THE INSTRUCTION MANUAL FOR EVERY LIVING ORGANISM ON EARTH WAS ONE OF THE GREATEST BREAKTHROUGHS IN THE HISTORY OF SCIENCE.

people. Everyone's life story is unique, including the life stories of the people who figured out that DNA contains that instruction manual and recognized the chemical code in which those instructions are written.

Unraveling DNA's secrets required the effort of many scientists working with many different kinds of equipment in many different laboratories. No two of them thought alike, their personalities and abilities were very different, and not all of them got along well with each other.

Some got more credit than others for breakthrough ideas and revolutionary discoveries that led to an understanding of DNA's spiral staircase structure—a shape called a double helix—and the way it functions as an organism's instruction manual. Some were honored and some were almost overlooked by history.

The two names best known for the double helix breakthrough are James Watson (1928–) and Francis Crick (1916–2004). But they could never have made their discoveries without the work of two other great scientists, Maurice Wilkins (1916–2004) and Rosalind Franklin (1920–1958).

The story of their complex personal interactions is as fascinating as the story of their quest for the secrets of DNA. Both of those stories—the scientific and the human—begin with the arrival of two brilliant but very different young scientists at two of England's most important laboratories in the early 1950s.

CHAPTER ONE

FOUR PEOPLE, TWO LABORATORIES, ONE GOAL

When the young American James Watson arrived at Cambridge University's Cavendish Laboratory in 1951, he had every reason to expect that his future was bright. One of England's two greatest universities, Cambridge, about a 60-mile (95-kilometer) drive north of London, had been the home to many of history's greatest scientists, including Sir Isaac Newton (1642–1727). The Cavendish had been

JAMES WATSON, SHOWN HERE AS A YOUNG PROFESSOR AT HARVARD UNIVERSITY, SEEMED DESTINED FOR GREATNESS FROM AN EARLY AGE. HE ENTERED COLLEGE AT AGE FIFTEEN AND COMPLETED HIS DOCTORATE AT TWENTY-TWO, AFTER WHICH HE JOINED CAMBRIDGE UNIVERSITY'S RENOWNED CAVENDISH LABORATORY. ELEVEN YEARS LATER, WHEN THIS PHOTO WAS TAKEN, HE SHARED THE NOBEL PRIZE FOR MEDICINE FOR THE WORK HE DID THERE ON THE STRUCTURE AND FUNCTION OF DNA.

founded in 1874, and its list of directors and faculty already included numerous Nobel Prize winners, including its director, Sir Lawrence Bragg (1890–1971), who was astonishingly young when he shared the 1915 Nobel Prize in Physics with his father, Sir William Henry Bragg (1862–1942).

The Braggs won their Nobel for developing a technique called X-ray crystallography, in which X-rays passed through solids and produced images that revealed the way the atoms within those solids were arranged. Now the Cavendish was using similar methods to look at the structure of large organic molecules. And Watson, who had entered college at age fifteen and completed his doctorate by twenty-two, was being invited to join that prestigious laboratory. There he would meet and work with Francis Crick, who was studying proteins. Within two years, their collaboration would lead to one of the greatest scientific breakthroughs of all time: the structure and function of DNA.

MEANWHILE IN LONDON

The same year as Watson arrived in Cambridge, Rosalind Franklin was returning to King's College in her hometown of London after three years in France, where she became known for her expert X-ray crystallography studies of the carbon in coal. Although King's was not as prestigious as Cambridge, it had a tradition

X-RAY CRYSTALLOGRAPHY

Most people know about X-rays from medical and dental images. Those images are similar to photographs because they show the exact structure of what the doctors or dentists are looking at. The images produced by X-ray crystallography are very different.

X-rays and light are both forms of electromagnetic waves, but X-rays carry more energy and can pass through body tissues rather than reflect from them to produce their images. They are considered electromagnetic waves because they are made of a repeating pattern of up-and-down or left-and-right electric and magnetic fields that travel through space.

Crystallographers use X-rays in a different way than doctors and dentists. Most solids are crystals, meaning that their atoms or molecules are in definite patterns. If you could look inside a crystal, you would see atoms or molecules arranged in evenly spaced layers, or planes, which lie in different directions. When X-rays strike a crystal, they reflect off each layer.

In Crick and Watson's time, the reflections from all of those planes would combine at a photographic plate. Today, scientists use arrays of tiny detectors instead of photographic film. Where reflected X-ray waves meet with their peaks in the same direction, they produce a maximum in intensity (or a dark spot on a photograph). Where the reflected waves meet with peaks in opposite directions, they cancel out and produce nothing. That process, called X-ray diffraction, produces a pattern of spots that scientists can analyze to determine the arrangement and spacing of the atoms or molecules in the crystal.

of excellence in experimental science that predated the Cavendish by forty-three years.

Franklin was hired to work in the laboratory of John Turton Randall (1905–1984), who had studied physics at the University of Manchester, where Lawrence Bragg was then department head. Much of Randall's work was funded by the United Kingdom's Medical Research Council (MRC). By the time Franklin arrived, Maurice Wilkins had been working with Randall for several years.

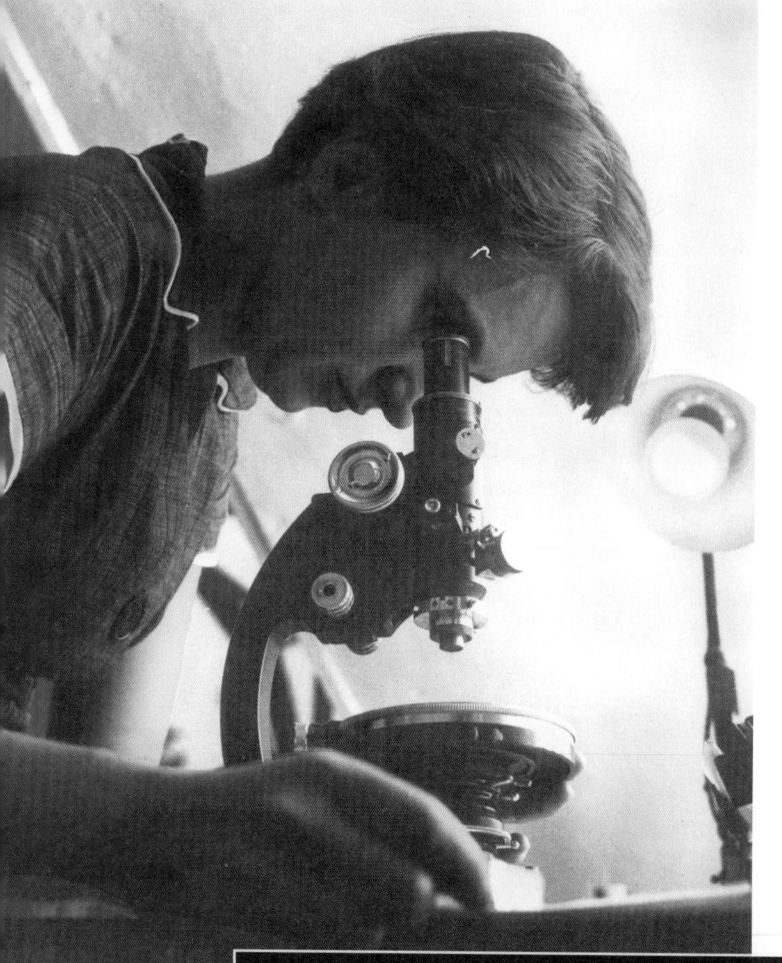

ROSALIND FRANKLIN ARRIVED AT KING'S COLLEGE IN LONDON THE SAME YEAR AS WATSON ARRIVED AT CAMBRIDGE. HER EXPERTISE IN PRODUCING AND ANALYZING X-RAY DIFFRACTION IMAGES OF LARGE MOLECULES LED TO THE BREAKTHROUGH DISCOVERY OF THE STRUCTURE OF DNA. HER EARLY DEATH MAY HAVE DEPRIVED HER OF THE NOBEL PRIZE SHARED BY WATSON, CRICK, AND WILKINS.

For the first few months, Franklin and Wilkins worked separately to learn more about a long molecule called DNA that made up the nucleus of living cells. They were cordial but not friendly. As time passed, despite a high regard for each other's work, the two clashed. Perhaps it was because Franklin loved a good argument and Wilkins shied away from disagreement.

No matter the reason, they fell into a pattern of speaking to one another only when necessary for the good of their research. Despite that frosty relationship, their work would ultimately prove to be of major importance to Crick and Watson's revolutionary discovery.

FRANCIS CRICK

To understand the human stories behind the DNA breakthrough, it is useful to look at the path that brought each of those four scientists to their two different laboratories. We begin with Francis Crick.

Francis Harry Crick was born at home in Northhampton, England, on June 8, 1916. Francis was a fair-haired, blue-eyed child with an insatiable curiosity. He was especially fascinated by anything having to do with science. After graduating high school at age eighteen, Crick entered University College in London, where he earned a bachelor of science degree in physics 1937. He remained there for graduate school but found his assigned research to be quite dull—determining the viscosity (the resistance to flow) of water at high temperature and pressure. He had almost finished his project when World War II broke out in 1939.

During the war, Crick worked on a project for the British navy. But when it ended he wanted to work on something new. As he relates in *What Mad Pursuit*, he applied "The Gossip Test." He paid attention to the subjects he most liked to gossip about with his friends. And what he liked to talk about the most was the borderline between the living and the nonliving, a science we know today as molecular biology.

Molecular biology is the study of the large molecules that make life possible. When Crick entered the field, the molecules of interest included proteins, such as hemoglobin, and DNA, which was known to be present in the cells of every living organism and to play a role in how inherited traits are passed from parent to offspring.

Although Crick used many of the same skills in

AFTER WORKING ON A PROJECT FOR THE BRITISH NAVY DURING WORLD WAR II, PHYSICIST FRANCIS CRICK WENT SEARCHING FOR A NEW PROJECT. HE WAS FASCINATED BY THE BORDERLINE BETWEEN LIVING AND NONLIVING SUBSTANCES, SO HE WENT TO WORK IN THE FIELD OF MOLECULAR BIOLOGY, USING X-RAY CRYSTALLOGRAPHY TO STUDY THE STRUCTURE OF LARGE MOLECULES. IN 1949, HE JOINED THE CAVENDISH LABORATORY, WHERE HE BEGAN TO INVESTIGATE PROTEINS.

molecular biology as he had used in physics, his transition still required a dramatic change in approach. Measurement in physics is usually precise and guided by clear mathematical laws and logic. Materials from living organisms are variable and complex. Crick had to learn about many new areas and techniques in biology and organic chemistry. He also learned how to apply X-ray crystallography to biological materials.

In 1949, after two years at Strangeways Laboratory in Cambridge, he joined the Cavendish. There, he began to investigate the structure of protein molecules.

JAMES WATSON

Francis Crick's wife, Odile, met James Watson before her husband did, and he intrigued her for an unusual reason. As Crick wrote in *What Mad Pursuit*, Odile described the young American with these words: "You know what—he had no hair!" What she meant was that he had an American crew cut, a hairstyle that was considered quite strange in Cambridge at the time.

James Dewey Watson was born in Chicago on April 6, 1928, and he quickly displayed remarkable intelligence. By the age of twelve, the young Watson was appearing on a popular radio show called *Quiz Kids*, where he and other extremely bright youngsters amazed listeners by answering difficult questions that most adults couldn't. At age fifteen, he graduated from high school and entered the University of Chicago. An avid birdwatcher and fascinated by the phenomenon of bird migration, he decided to major in zoology.

It was around that time that a book changed Watson's interests in a fateful direction. In *What Is Life?*, theoretical physicist Erwin Schrödinger proposed that to understand life, it would be necessary to figure out how genes work. Watson was determined to do just that.

After graduating from Chicago and earning a Ph.D. in zoology from the University of Indiana in 1950, Watson went to Copenhagen, Denmark, to work with a biochemist there, but found he didn't like the work at all.

At first he wanted to return to the topic of his Ph.D. research. Then he heard a talk by Maurice Wilkins, an expert in X-ray crystallography, who showed a slide of an X-ray diffraction picture of DNA. Wilkins's talk inspired Watson. He applied for a position at the Cavendish Laboratories and was hired. Sir Lawrence Bragg brought him on to work with Max Perutz, another X-ray crystallographer, to study the structure of hemoglobin. But it was not long before he and Crick teamed up to study DNA.

MAURICE WILKINS

Maurice Hugh Frederick Wilkins was born in Pongaroa, New Zealand, of Irish parents on December 15, 1916. The family moved to the New Zealand capital, Wellington, when he was a baby, and then to Birmingham, England, when he was six.

After completing his high school education in Birmingham, Wilkins studied physics at St. John's College of Cambridge University. After earning his bachelor's degree in 1938, he remained at Cambridge for his graduate education, where John Randall supervised his Ph.D. research on the glow-in-the-dark property of phosphorescence, which he completed in 1940.

In 1945, after completing wartime work on radar screens in England and the nuclear fuel for the atomic bomb in the United States, Wilkins rejoined Randall, who was then chair of the physics department at St. Andrews University in Scotland. Randall was hoping to get funds from the Medical Research Council to create a laboratory where methods of physics, such as X-ray crystallography and advanced microscopy, could be

MAURICE WILKINS WAS WELL ESTABLISHED IN THE FIELD OF X-RAY CRYSTALLOGRAPHY AT KING'S COLLEGE AND HAD ALREADY BEGUN TO STUDY DNA WITH ADVANCED EQUIPMENT WHEN ROSALIND FRANKLIN ARRIVED THERE IN JANUARY 1951. HE IS SHOWN HERE WITH A MODEL OF THE DNA MOLECULE IN 1962, THE YEAR HE SHARED THE NOBEL PRIZE FOR MEDICINE.

applied to biology. The idea of a biophysics laboratory was new and appealing, but the MRC said that it belonged in a different university.

The next year, Randall was hired as the Wheatstone Professor of Physics at King's College in London, which put him in charge of the department as well. The MRC provided funds to start a biophysics unit, and Wilkins became its assistant director. King's soon became one of the leading research institutions in applying X-ray diffraction and crystallography to DNA. In the summer of 1950, Wilkins had ordered new and better X-ray equipment to study samples of DNA that the college had received from a noted Swiss scientist.

Although neither Wilkins nor King's College could have anticipated the scientific and human drama that would unfold because of that equipment, they were poised to become key players in a scientific revolution that would answer Schrödinger's question, *What Is Life?*

ROSALIND FRANKLIN

Also in the summer of 1950, Randall had invited an expert X-ray crystallographer named Rosalind Franklin to join the department for a three-year fellowship to study the structure of proteins. She agreed to come after she finished her work at a French government laboratory in Paris. She finally arrived on January 5, 1951. By then, her assignment had changed. Instead of studying proteins, she would be studying DNA.

Rosalind Elsie Franklin was born in London on July 25, 1920, to a wealthy and socially prominent Jewish family. Even as a small child, Rosalind showed a sharp and unusual mind. After a family summer vacation, her aunt Mamie Bentwich, in a letter to her husband, Norman, who was Britain's attorney general to Palestine, described the six-year-old Rosalind as "alarmingly clever—she spends all her time doing arithmetic for pleasure, & invariably gets her sums right."

Aunt Mamie used those words with affection, but there was reason to be alarmed by Franklin's intelligence. The bright young women of her social class were expected to get an education but not to have careers. They were expected to raise families and take the lead in charitable and social causes within the community.

Franklin's passion for math and science, which grew stronger as she moved toward adulthood, moved her in a very different direction. She was a strong-willed child who quickly learned to speak her mind. That frequently led to arguments with her friends and family. Most of the people who knew her recognized that she was not being intentionally disagreeable. She argued simply because she enjoyed the mental tussling.

Franklin's early education was at the exclusive St. Paul's Girls' School in London. In 1938, she enrolled in the women-only Newnham College of Cambridge University. Although the women graduates of Newnham could not earn a full Cambridge degree, they were still able to take full advantage of the university's excellent courses. Franklin excelled in her studies and laboratory work.

Unlike many of her friends who served in the women's army service during the war, Franklin found herself studying the physics and chemistry of coal. Her excellent research led to a Ph.D. degree from Cambridge. She was now Dr. Franklin, a young experimental scientist whose work could open many doors.

In 1947, Franklin got what was then the job of her dreams, to continue research on the physics and chemistry of carbon in coal and graphite. And it couldn't have been in a better place—Paris at an outstanding government laboratory, Laboratoire Central des Services Chemiques de l'Etat. Since she loved outdoor activities, travel, and different cultures, her life could hardly have been better.

Then in the summer of 1950, Randall invited Franklin to return home to study proteins at King's College. Even though she loved living in France, the opportunity was too good to pass up.

SCIENCE IN THE 1950S

CHAPTER TWO

If you ask most Americans old enough to remember the 1950s, they are likely to recall a period of relative calm and comfort. But those memories leave out the controversies that also mark the time period. To appreciate the science of DNA, it is useful to understand the world in which it was done.

A FASCINATING AND TURBULENT DECADE

During the years of World War II (1939–1945), China and the Union of Soviet Socialist Republics (USSR), dominated by Russia, were allies with the United States, the United Kingdom, France, and most other western European nations against the Axis powers, led by Germany and Japan. But by

WHILE WATSON, CRICK, WILKINS, AND FRANKLIN STRUGGLED TO UNDERSTAND DNA, THE WORLD WAS ENGAGED IN A POLITICAL BATTLE BETWEEN TWO VERY DIFFERENT ECONOMIC SYSTEMS, COMMUNISM AND CAPITALISM. THE MAJOR COMMUNIST POWER WAS THE UNION OF SOVIET SOCIALIST REPUBLICS, MADE UP OF RUSSIA AND ITS SATELLITE NATIONS. EVERY YEAR ON THEIR NATIONAL MAY DAY HOLIDAY, THE SOVIETS WOULD DEMONSTRATE THEIR MILITARY MIGHT IN A PARADE THROUGH MOSCOW'S RED SQUARE.

the 1950s, new alliances and conflicts had developed. Now the major division was between communist and capitalist economic systems.

A Cold War developed between the West, led by the United States, and the communist world led by the USSR, which had been under communist rule since 1917. In 1949, the USSR was joined by "Red China," when a communist government took over the Chinese mainland.

The Cold War turned temporarily hot in 1950, when fighting broke out on the Korean peninsula. It lasted until a cease-fire agreement in 1953 and resulted in a country divided into the communist north and the capitalist south. More than a million soldiers, including almost thiry-seven thousand Americans, died in the conflict.

Meanwhile, across Eastern Europe, communist governments with strong Russian influence rose to power. While not openly at war, the two sides were in a life-or-death competition over world politics, ideology, economics, and military strength.

Even against this backdrop of world struggle, Americans at home were feeling confident enough to marry and raise families. There were worries about inflation and labor struggles, but overall America was booming, and so was the birth rate.

It was also a time of great scientific and technological advances. People were living longer and healthier lives because of advances in medicine. One of the most important was the development of a vaccine to protect against the often crippling, sometimes fatal disease of poliomyelitis, commonly called polio. A mass inoculation campaign began almost immediately after the announcement that the vaccine was safe and effective in April 1955.

The invention of transistors in 1957 allowed radios to shrink to a handheld size and hearing aids that could fit inside an ear. The credit card surged into widespread use for the first time. Most homes had black and white televisions. And UNIVAC, the first commercial computer, was a sign of great changes ahead.

One technology overshadowed all others in the Cold War years: nuclear weapons. The United States had ended World War II by exploding two atomic bombs over Japan. These used a form of energy created when the tiny central parts of atoms called nuclei (singular: nucleus) split apart in a process called fission. No other country had an "A-bomb" until the Soviets tested their first fission device in September 1949.

With that, the race was on to develop an even more powerful weapon, the "H-bomb," based on the fusion (combination) of hydrogen nuclei to create helium nuclei and release energy in the same way the sun creates its light and heat. The United States tested its first H-bomb in November 1952. The Soviets again followed suit by exploding an H-bomb in August 1953.

THE POSITIVE SIDE OF NUCLEAR FISSION

Scientists and engineers also saw a positive side to nuclear fission. If it could be used in a controlled reactor, it could generate enormous amounts of electrical energy for both civilian and military use. The first fission-powered nuclear submarine was completed in 1953, and the first commercial electrical power plant began operation in 1957. Today, nuclear fission power plants supply nearly 20 percent of the electricity in the United States and more than 12 percent worldwide.

Still the science of the 1950s transformed twentieth-century technology in ways that were mostly beneficial. And the discovery of the structure and function of DNA during that same decade promises to produce changes in twenty-first-century life and medicine that go far beyond anything that Crick, Watson, Wilkins, and Franklin could imagine.

EVOLUTION, INHERITANCE, GENES, AND DNA

Like all other great scientists, Crick, Watson, Wilkins, and Franklin understood that they were building on a foundation established by others who came before them. The base of that foundation is the idea that all life on Earth traces back billions of years to a time when ordinary chemistry produced simple structures that could reproduce themselves. Those structures were the first living organisms. They reproduced, but they could also change slightly.

To get from the first primitive organisms to today's world filled with many species, nature followed a process that Charles Darwin (1809–1882) called evolution by natural selection. His book *On the Origin of Species*, published in 1859, describes the process by which traits are passed down from one generation to the next, but it does not include the fact that those traits are carried within the organism on units that we now call genes.

That discovery first came from an Austrian monk named Gregor Mendel (1822–1884). In 1856, he began experiments with peas that traced how physical traits are passed from one generation to the next. He completed his experiments in 1863 and published his results, which included a discussion of statistics, in a relatively unknown scientific journal in 1866. Mendel's results were almost lost to history but were rediscovered around 1900, and scientists quickly realized their importance.

As scientists observed how cells divided, they realized that the genes had to be carried on a substance within the cells. But what was that substance? An early step toward answering that question came in 1869 from a Swiss biochemist, Friedrich Miescher (1844–1895), who was instrumental in

IN THE MID-NINETEENTH CENTURY, CHARLES DARWIN LAUNCHED A REVOLUTION IN THE LIFE SCIENCES WITH HIS THEORY OF EVOLUTION BY NATURAL SELECTION. DARWIN'S THEORY STATES THAT SPECIES CHANGE OVER TIME BY DEVELOPING NEW TRAITS THAT ENABLE THEM TO ADAPT TO THEIR ENVIRONMENTS AND THEN PASSING THOSE TRAITS TO THEIR OFFSPRING. BUT HIS THEORY DID NOT ADDRESS WHAT CAUSES THOSE TRAITS. WE NOW KNOW THAT GENES WITHIN AN ORGANISM'S DNA CARRY INSTRUCTIONS FOR BUILDING THAT ORGANISM.

isolating a new molecule, nuclein, from the nucleus of a cell. Though he didn't know it, what he had found was DNA.

It was a large molecule, but not nearly as large and complex as proteins. Miescher suspected that nuclein might be involved in genetic inheritance, but most other scientists were convinced that the more complex proteins were responsible for the genetic code. It was seventy-five years before this view was changed.

In 1882, a German anatomist named Walther Flemming (1843–1905) was observing the process in which one cell divides into two. He noticed that the nuclein became a thread made up of several individual units, which became known as chromosomes. After cell division, the chromosomes in each of the new cells are exactly the same as the chromosomes of the original. That suggested to scientists that chromosomes might be responsible for carrying the traits from one generation to the next.

That idea gained support from the work of American biologist Thomas Hunt Morgan (1866–1945). Starting in 1907, Morgan bred millions of fruit flies and documented how mutant traits were inherited and passed on. He showed that genes for those traits were located on the chromosomes, a discovery that earned him the Nobel Prize in Physiology or Medicine in 1933.

The final step in showing that DNA carried the genetic code came in 1944 from Oswald Avery

CRICK AND WATSON COMBINED CHEMICAL CLUES WITH THE X-RAY CRYSTALLOGRAPHY OF WILKINS AND FRANKLIN TO MAKE THEIR BREAKTHROUGH DISCOVERY. THEY ARE SHOWN HERE IN 1953 NOT LONG AFTER PUBLISHING THEIR FINDING THAT DNA MOLECULES HAD THE SHAPE OF A DOUBLE HELIX WITH ITS BACKBONES MADE OF SUGARS AND PHOSPHATE, JOINED BY CHEMICAL BASES THAT CARRIED THE CODE OF LIFE.

(1877–1955), Maclyn McCarty (1911–2005), and Colin MacLeod (1909–1972). By transferring DNA from one pneumonia bacterium into another, they showed that the second bacterium developed the exact genetic traits of the first.

But what was it about DNA that enabled it to hold genes, and how did those genes function in the bodies of living organisms? The keys to answering those questions would come from the research of Crick and Watson at the Cavendish Laboratory and Wilkins and Franklin at King's College.

CHAPTER THREE

THE DNA RACE BEGINS

Scientific research is known for its cooperative spirit. New knowledge develops from sharing results of many different investigations. But scientists are also human beings who compete with one another for glory. In their world, that glory comes from being the first to make an important discovery.

Those two sides of science were both on display as the quest for the secrets of DNA became the hottest area in biology. Crick, Watson, Wilkins, and Franklin became key players in a race for scientific acclaim.

PAULING SETS THE PACE

In April 1951, Linus Pauling (1901–1994), an American chemist, along with Robert Corey

(1897–1971) and Herman Branson (1914–1995), published their findings on what was called the a-helix, which formed the backbones of tens of thousands of proteins. In mathematics, a helix follows the shape of a coil, like the threads of a screw, a spring, or a spiral binding of a notebook.

Could DNA also have a helical shape? If so, how did that shape allow it to carry the genetic code? Those questions galvanized the highly competitive, but unknown, Watson who was determined to arrive at the answer to DNA before his much more famous American colleague, Pauling, did.

Pauling influenced both Watson and Crick in another critical way. Besides using X-ray images of proteins, Pauling also built physical models to work out his discovery. It was important to know exactly how all the parts of a protein fit together with their helical backbone. Watson and Crick used the same method to arrive at theirs.

EVEN THOUGH SCIENTISTS VALUE COOPERATION, THEY CAN ALSO BE VERY COMPETITIVE, ESPECIALLY WHEN IT COMES TO BEING THE FIRST TO PUBLISH AN IMPORTANT RESULT. IN THE RACE TO DISCOVER THE STRUCTURE OF THE DNA MOLECULE, AN AMERICAN TEAM LED BY LINUS PAULING, WAS THE EARLY FAVORITE. WATSON AND CRICK LEARNED THE IMPORTANT TECHNIQUE OF BUILDING CHEMICAL MODELS FROM PAULING'S WORK.

A FAST FRIENDSHIP

When people speak of a fast friendship, they are using "fast" to mean having a strong bond, as in the word "fasten." Crick and Watson's friendship was fast in both senses of the word. It did not take long for them to bond both personally and professionally.

A few days after arriving in Cambridge in mid-October 1951, Watson met Crick at the Cavendish Laboratory. Each man immediately sparked to the intellectual brilliance of the other.

In *The Double Helix*, Watson says that he "immediately discovered the fun of talking to Francis Crick." Crick concurs in *What Mad Pursuit*, saying, "Jim and I hit it off immediately, partly because our interests were astonishing similar and partly, I suspect, because a certain youthful arrogance, a ruthlessness, and an impatience with sloppy thinking came naturally to both of us."

They were soon sharing lunches together at the local pub so that they could continue their conversations. They spent so much time talking that when a vacant office became available at the Cavendish, the other scientists gladly gave it to Watson and Crick so that they could talk to one another without disturbing the rest of them.

Some of this disturbance stemmed from Crick's loud laugh, which could bring an entire room to a standstill, and his tendency to talk at great length and with boisterous enthusiasm about whatever had seized

THE REMARKABLE CAREER OF LINUS PAULING

With a career spanning from 1921 to 1994, Pauling was one of the world's most famous chemists and made many significant discoveries. He is best known for using the principles of quantum mechanics—one of the twentieth century's greatest breakthroughs in physics—to explain the various ways atoms bond together to form molecules. For that work, he was awarded the 1954 Nobel Prize in Chemistry. Throughout his long life, he was a tireless speaker against war and for abolishing atomic bomb tests. For that effort he won the Nobel Peace Prize for 1962. He is one of four people to win two Nobel Prizes and the only person to win two unshared Nobels.

his interest at the moment. Yet Crick thought it was Watson that was the more outspoken of the two, possibly because as an American, Watson was more used to speaking his mind.

When they met that fall, neither one of them was officially sanctioned to work on the problem of DNA. There was an unwritten but very powerful code of etiquette in British science that prevented one scientist from competing in an area that officially belonged to another scientist. This wasn't the case at all in America, where nothing was holding back Pauling from pursuing DNA, and the pair of fast friends knew it.

So Watson and Crick decided to divide their time between working on their official projects and solving

EVEN THOUGH WATSON AND CRICK WERE OFFICIALLY ASSIGNED TO OTHER PROJECTS, THEY RECOGNIZED THE IMPORTANCE OF THE RACE TO UNDERSTAND THE STRUCTURE AND FUNCTION OF DNA. WATSON'S ASSIGNMENT WAS TO WORK WITH MAX PERUTZ TO CRYSTALLIZE MYOGLOBIN, A MODEL OF WHICH IS SHOWN HERE.

the DNA mystery. Crick was meant to be finishing his Ph.D. thesis on the X-ray diffraction of proteins. Watson was supposed to be helping Max Perutz crystallize myoglobin (a protein found in muscle cells) to use in X-ray pictures. Those projects were too important to neglect, but they were not as urgent as the race to understand the shape and function of DNA.

A DIFFICULT PAIRING

In a marked contrast between the close Crick-Watson working relationship at the Cavendish, Maurice Wilkins and Rosalind Franklin, though professional colleagues at King's College, had formal and often strained interactions.

Francis Crick was an old friend to Wilkins, whom Watson was eager to know better, especially since it was a lecture by Wilkins that inspired Watson to get into X-ray crystallography. Wilkins's X-ray images were crucial to figuring out what the structure of DNA might be, and, unlike Crick and Watson, his official work was to study that intriguing molecule. The addition of Franklin's expertise was sure to be valuable in that work.

Unfortunately, Wilkins and Franklin never got along. For the first few months their relationship was formal but cordial, but as time passed, their personalities began to clash. Franklin was a brilliant scientist, but in the male-dominated world of science at the time, she had tough going. She wasn't even allowed to have coffee in one of the faculty rooms reserved for men only.

From Watson's perspective, Franklin was extremely defensive about her position and secretive about her work because she wanted to make sure everything she did was beyond criticism. Franklin's biographers Anne Sayre and Brenda Maddox see things differently. They agree that, as Watson noted, Franklin was a perfectionist about details. But she could handle criticism as well as any scientist. In fact, she relished a good argument, whether it was about politics or how to interpret scientific observations and data.

Franklin's willingness to argue and reluctance to share incomplete results may have been particularly disagreeable to Wilkins because he had many more years of experience. He may have expected her to behave more like his assistant, but she was officially assigned to other projects. It made for a tense situation. It also meant that Watson and Crick weren't able to see Franklin's new X-ray photos until she was satisfied with their quality.

EARLY RESULTS

In November 1951, Crick and a colleague, Bill Cochrane (1922–2003), worked out a significant mathematical problem on how to interpret X-ray diffraction photographs of helical molecules. It helped overcome some of the problems posed by the fact that an X-ray photo was giving only a two-dimensional impression of a three-dimensional atomic structure. Even with this advancement, it was impossible to be precisely sure of the complete structure from only an X-ray photo. Much of it was a matter of interpretation and figuring out atomic measurements.

AS THEY PUT TOGETHER EVIDENCE OF DNA'S HELICAL STRUCTURE, WATSON AND CRICK DESIGNED AND BUILT PHYSICAL MODELS OF THE MOLECULE TO TEST THEIR IDEAS OF HOW ITS VARIOUS PIECES—THE SUGAR-PHOSPHATE BACKBONE AND THE FOUR CHEMICAL BASES, A, G, T, AND C—FIT TOGETHER.

Watson and Crick decided to start building some physical models, confident that putting a model together would help them get the answer in a short time. They started with several pieces of established information:

- They learned from Wilkins that the thick diameter of DNA meant it had to consist of more than one chain. The trouble was that it might be anywhere from two to a hundred chains.
- They knew it had to have a "backbone" that was a sugar-phosphate combination, but they didn't know if this backbone was a central axis or on the outside of the structure.
- There were four chemical bases that were the prime components of DNA. The four bases are adenine, guanine, thymine and cytosine. They are referred to by their first initials: A, G, T, and C. What no one knew yet was how these four bases fit into the DNA molecule or how they could function to carry genetic information.

In November 1951, Watson went to London to hear Wilkins and Franklin speak about their work. Franklin showed slides of two forms of protein crystal X-ray photos: the A, or "dry" form; and the B, or "wet" form. When protein crystals dry out, they contract into tight, jumbled forms. The B form, which held more water, was harder to achieve but gave a simpler and more revealing photo.

Unfortunately, Watson didn't take notes and didn't yet know much about X-ray crystallography, so the information he gave to Crick upon his return to Cambridge wasn't entirely accurate. Crick deduced that their choices were narrowed down to a two-, three-, or four-part chain. They also needed other information, which they found by poring over a book by Pauling, *The Nature of the Chemical Bond*. They used it so much that they had to buy two copies, one for each of them.

Next they started to build their model of DNA. They took some models of carbon atoms that they already had. To some of those, Watson added bits of copper wire to convert them into larger-sized phosphorous atoms.

Based on the slightly faulty information from Watson's memory, he and Crick built their first model consisting of three chains in a triple helix with the phosphates on the inside of the structure. Wilkins and Franklin traveled to Cambridge to see it, but Franklin gave them a devastating critique, pointing out everything they'd gotten wrong.

Worse yet, the head of the Cavendish, Sir Lawrence Bragg, discovered what was going on and ordered Watson and Crick to confine themselves to the work they were supposed to be doing. Watson and Crick sent the parts for their model to the King's group, but they were never used. The year 1951 ended with the pair at a low ebb.

STAYING OUT OF SIGHT

For the first half of 1952, Watson and Crick adopted a different strategy. They would lie low to appease Bragg and keep doing what he was paying them for. But they did not let the DNA challenge get far from their minds. They needed new data before they could make another attempt to solve the problem, so they continued to read Pauling's book and do quiet study. To keep their discussions of DNA away from working hours, they restricted them to lunchtimes at the Eagle, the local pub.

Their reading brought them an important piece of the puzzle, which came from the research of Austrian-born biochemist Erwin Chargraff (1905–2002) and his students at Columbia University. After many painstaking analyses of DNA samples beginning in 1949, Chargraff's team found that there was always nearly the same amount of A as there was T, and the same amount of G as there was C. The amount of the four bases would vary from one species to another, but the amounts of A and T were always the same, as were the amounts of C and G. These became known as Chargraff's rules.

Chargraff himself visited Cambridge during Crick and Watson's lying-low period. He met the two unknowns at a dinner and was not favorably impressed. He didn't like Watson's American brashness, and he thought Crick was selling an idea that he didn't really understand—which was reinforced by Crick's inability to remember the chemical differences between the four bases.

Even though the dinner did not go well, Crick and Watson got more out of it than Chargraff. Chargraff's name may have been attached to his famous results, but it was ultimately Crick and Watson who figured out what Chargraff's rules meant.

It was Crick who first became convinced that Chargraff's rules were the key to understanding the structure and function of DNA. For a time, he had trouble getting Watson to agree with him. Summer came and went with no advances.

Meanwhile, the souring relationship between Wilkins and Franklin at King's College was slowing Crick and Watson's progress on DNA. Franklin was becoming convinced that the sugar-phosphate backbone was on the outside of the molecule. But Wilkins backed away from doing anything that would antagonize her, so he was unable to show her evidence to his friends at the Cavendish. Without access to her pictures, Watson and Crick were unable to pursue that line of investigation.

PAULING GETS IT WRONG

In September 1952, Linus Pauling's son, Peter, arrived to work at the Cavendish and shared offices with Watson and Crick. Their spirits hit an all-time low when Peter got a letter in December stating that his famous father had figured out a structure for DNA. It seemed the race was over. But in fact it was merely the end of the first lap.

They waited for months for some dramatic announcement out of Pauling. As 1952 turned into 1953, Franklin also announced a change—she was leaving King's and giving up her study of DNA. Wilkins planned to get back to his study of DNA as soon as she left.

In January 1953, Watson obtained an advance copy of Linus Pauling's paper on the structure of DNA from Peter. The elder Pauling had decided upon a three-chain helix with the sugar-phosphate backbone in the center.

Watson and Crick instantly knew that Pauling had somehow made a major mistake and not caught it. They knew this structure was wrong. They also knew that as soon as the paper was published, Pauling would learn of his mistake and probably turn around to find the right solution.

Watson and Crick knew the race was on again. They plunged back into the problem with renewed energy. Time was short, since Pauling was sure to learn about his error soon. But for now, they were in the lead.

CHAPTER FOUR

SEEING THE DOUBLE HELIX

With no time to lose, Watson headed to King's College. He tried to talk to Franklin about Pauling's paper, but they ended up having a terrible argument. Afterward, Wilkins, in a moment of sympathy, showed Watson something of profound significance—it was the now-famous Photo 51, an X-ray photo of the B form of DNA that Franklin had taken months earlier.

THIS IMAGE IS THE NOW FAMOUS PHOTO 51, FRANKLIN'S CAREFULLY PRODUCED X-RAY DIFFRACTION IMAGE OF THE WET OR "B" FORM OF DNA. EVEN THOUGH FRANKLIN WAS NOT READY TO SHARE IT, WILKINS SHOWED IT TO WATSON, WHO IMMEDIATELY RECOGNIZED WHAT IT MEANT. HE RETURNED TO CAMBRIDGE AND PROMPTLY WENT TO THE MACHINE SHOP TO ASK THE MACHINISTS FOR THE PARTS HE NEEDED TO BUILD A MODEL OF THE DOUBLE HELIX.

It was not Wilkins's place to reveal the photo. Franklin had not yet published it: she was still trying to figure out exactly what it showed. To Watson, it immediately signaled a new direction. In *The Double Helix*, he wrote, "The instant I saw the picture my mouth fell open and my pulse began to race. The pattern was unbelievably simpler than those obtained previously. Moreover, the black cross of reflections which dominated the picture could arise only from a helical structure."

PUTTING IT ALL TOGETHER

On the train back to Cambridge, Watson sketched out the image as best as he could remember it on the margin of a newspaper. He decided it was showing a two-chain structure, a double helix like a spiral staircase with two rails.

Watson went to Sir Lawrence Bragg and convinced him that Cavendish should try to beat Pauling to the punch. Bragg authorized Watson to start building models again. Watson dashed straight to the machine shop to get them started making parts, hoping to have them within a week.

Later in the day, Watson finally connected with Crick at the office. "Reporting that even a former bird-watcher could now solve DNA was not the way to greet a friend bearing a slight hangover," Watson wrote in *The Double Helix*.

Crick wasn't ready to accept a two-chain concept just yet. He wanted them to keep open minds as they worked on the models. When some of the parts arrived a few days later, Watson experimented with a two-chain model that had the sugar-phosphate backbone on the inside.

Since he was still waiting for some of the other parts, and since he had nothing to lose, Watson also began building a different version of the model with the backbone on the outside of the molecule, as Franklin had suggested. At first, Watson thought that the pairing of bases was like-to-like, meaning that A would always pair to A, and G would always pair to G, etc.

Jerry Donohue, a visiting American scientist, helped with another important piece of information. Watson and Crick were using forms of the four bases that were commonly found in textbooks. Donohue pointed out that new thinking said those forms were wrong, and he gave Watson the new forms that he should be using. The small but vital changes in the shapes of the bases made a big difference in how the model could be put together. This instantly did away with Watson's like-to-like theory. Furthermore, Crick kept insisting that they had to keep Chargraff's rule in mind.

By late February, the machined parts had not arrived. Impatient with waiting, Watson cut out cardboard pieces to use until the metal parts arrived. On February 28, 1953, Watson cleared a space off his desk

DNA AND CELL DIVISION

The double helix structure of DNA is like a spiral ladder with rungs connecting two rails. The rails are the sugar-phosphate backbones that Watson and Crick now realized were on the outside. Each rung is a pair of bases, either A and T or C and G. A base combined with sugar and phosphate molecules makes up one of the nucleotides that Crick and Watson would later find to carry the genetic code.

When a cell divides, the rungs of the ladder split like the halves of a zipper. Each half is then a sequence of nucleotides. The surrounding body fluid contains all the chemicals needed to build more DNA, so each base on the two halves of the split ladder attracts its matching base from the fluid. Every A finds a T, every C finds a G, every G finds a C, and every T finds an A. That completes two new sets of rungs connected to single helixes. The open ends of the rungs attract the sugar and phosphate molecules needed to create the second helix. That completes the formation of two new DNA molecules with the same sequence of nucleotides as the original.

and began playing around with his cardboard pieces. A friend interrupted him for a few minutes.

When Watson returned to his work, his eye fell on a certain combination of the cardboard pieces. In *The Double Helix*, he relates the moment when he suddenly "became aware that an adenine-thymine pair held together by two hydrogen bonds was identical in shape to a guanine-cystosine pair held together by at least two hydrogen bonds."

The combination of A=T and G=C immediately brought to mind Chargraff's rule.

Crick was only halfway through the door when Watson told him they'd found the answer. Though Crick was cautious at first, remembering their previous mistakes, it only took him a few minutes to see that it worked.

Not only did it give them a double helix whose shape and measurements fit the known data, it also gave them a mechanism for a genetic code that could easily create a template and reproduce itself. (See the "DNA and Cell Division" sidebar.) The only thing that remained was to get the proper metal parts to build a real model and confirm their find.

TELLING THE WORLD

All the same, Watson relates in *The Double Helix*, that "...I felt slightly queasy when at lunch Francis winged into the Eagle to tell everyone within hearing distance that we had found the secret of life."

Crick also gave the news to his wife, Odile, at dinner. In *What Mad Pursuit*, Crick reveals that, years later, Odile confessed she hadn't believed him. "You were always coming home and saying things like that," she said, "so naturally I thought nothing of it."

The next day, Watson quickly managed to get the parts from the machine shop. Since only one person at a time could work on the model, Watson put the first arrangements together with Crick checking it afterward. The final model featured ten base pairs and was 6 feet (2 meters) high.

Bragg was instantly enthusiastic when the saw the model, but in good scientific form he insisted on having it double-checked by an expert in organic chemistry. The expert approved it. Next, Wilkins saw the model and checked its measurements against his own data. Both Franklin and he said their data supported the model. It was agreed that

THIS RECONSTRUCTION INCLUDES SOME OF THE METAL PLATES FROM CRICK AND WATSON'S ORIGINAL DNA MODEL. IT REPRESENTS ONE COMPLETE TURN OF THE HELIX. THE LABELS A, T, C, AND G ARE VISIBLE ON THE PLATES THAT REPRESENT THE BASES. WIRES REPRESENT THE BONDS THAT FORM THE SUGAR-PHOSPHATE BACKBONE AND CONNECT IT TO THE BASES.

Wilkins, Franklin, and their colleagues at King's would write a paper about their X-ray results to go out at the same time as the paper that Watson and Crick would write about their DNA discovery.

Watson got his sister to type up the manuscript, and Odile Crick drew the picture of the double helix. Watson wanted to be cautious about what they put in the paper, but Crick insisted upon one line indicating that they had realized the potential of the double helix structure to be self-replicating. That line read, "It has not escaped our notice that the specific pairing we have postulated immediately suggests a possible copying mechanism for the genetic material." This was possibly the understatement of the century.

Bragg approved it, and it was sent to the prestigious scientific journal *Nature*, where it was promptly accepted for publication. On April 25, 1953, the article appeared, entitled "A Structure for Deoxyribose Nucleic Acid." The authors flipped a coin to see whose name would come first. That's why the byline read Watson and Crick, rather than Crick and Watson.

The paper quickly earned them wide acclaim, but nowhere did it mention the role that Franklin's Photo 51 played in putting them on the right track.

CHAPTER FIVE

HOW DNA BUILDS ORGANISMS

The discovery that DNA contains the code for inheritance opened the door to decades of research, which is continuing today. Among the most important findings are how to read an organism's genetic code and how that code guides the development of that organism's unique traits.

As we learn more about the science of DNA, we also look for ways to apply it to improve human life. To understand those new applications, it is important to understand how DNA carries genes and guides the way that nature builds organisms.

SECRETS IN THE CELLS

Every organism on Earth is made up of cells. Some living creatures have only single cells, while

THIS SIMPLE SINGLE-CELLED LIFE-FORM IS A PARAMECIUM, A VERY PRIMITIVE ANIMAL. IT IS MADE UP OF A MEMBRANE THAT CONTAINS A JELLY-LIKE FLUID CALLED CYTOPLASM AND SEVERAL ORGANELLES, INCLUDING A NUCLEUS THAT HOLDS ITS DNA. ON THE OUTSIDE OF THE MEMBRANE ARE HAIR-LIKE CILIA THAT IT MOVES TO PROPEL ITSELF THROUGH WATER TO CAPTURE AND CONSUME BACTERIA.

others, like plants and animals (including humans), are made up of trillions of cells. In those complex life forms, cells are specialized to become particular parts, such as skin, hair, or teeth, or to form organs like the liver, heart, or brain. Yet inside each cell of each individual organism is an instruction book to build that entire individual. That instruction book is written in its DNA.

A living cell is a membrane surrounding a jelly-like fluid called cytoplasm. In life-forms classified as plants, animals, fungi, and protists, the cytoplasm surrounds organelles, which include the DNA-containing nucleus and other small "organs" that serve various functions (such as creating proteins and molecules that fuel the body). In bacteria and archaea, the cytoplasm has no organelles, but it does have a nucleoid region containing its unique DNA.

As Crick and Watson discovered, DNA is like a spiral ladder. If you follow one rail of the ladder, you will encounter a sequence of bases, each of which is either A, C, T, or G. That sequence is the genetic code, or genome, of the organism. When the cell divides, proteins surround its DNA to create a threadlike structure containing units called chromosomes. Within each chromosome are certain DNA sequences that make up the genes. There can be hundreds or thousands of genes on each chromosome. Chromosomes also contain DNA that does not seem to be instructions and is therefore not part of genes. This is called non-coding DNA (or sometimes "junk DNA"), and it makes up most of the molecule.

In *What Mad Pursuit*, Crick describes DNA as "a very long chemical message written in a four-letter language."

Here's another way to look at it:

Genome = the entire book of instructions for your body
DNA = the string of letters and punctuation marks that make up the book from beginning to end
Chromosomes = the chapters of the book
Genes = the readable sentences of each chapter
Non-coding DNA = letters between readable sentences

Each organism can be identified by its genome. All members of the same species have very similar genomes. Except for abnormal individuals, they all have the same number of chromosomes with the same kinds of genes on those chromosomes. A gene can exist in more than one variety called alleles. In some cases, a single gene affects a single trait such as eye color. A normal person can have blue eyes or brown eyes, and the difference comes from the particular alleles that person inherits from his or her parents.

Because the human genome contains a bit more than twenty thousand genes, and most genes have more than one normal allele, individual humans have quite a bit of variation in their genomes. But all that variation still represents only a small part of the whole

HOW CHROMOSOMES VARY

The number of chromosome pairs varies from organism to organism. Here are some examples:

- MOSQUITO = 6
- ONION = 16
- CAT = 38
- HUMAN = 46
- DOG = 78
- GOLDFISH = 94

genome. The individual genomes of any two people are 99.9 percent the same.

They are far more similar to each other than they are to a cat genome or a flower genome, which have different numbers of chromosomes and different genes on those chromosomes. Chromosomes come in pairs. Humans have twenty-three pairs or forty-six chromosomes altogether. A mother's egg and a father's sperm each carry one chromosome from each of that parent's pairs: twenty-three in the egg and twenty-three in the sperm. When they come together, the twenty-three chromosomes form the complete set of forty-six, and in this way each of us inherits characteristics from both of our parents.

READING THE GENETIC CODE

As Watson and Crick discovered, each "letter" of the genetic code can be one of four possible base pair combinations:

- A = T
- T = A
- G = C
- C = G

We now know that a single human gene can contain anywhere from twenty-seven thousand to two million of these base pairs. But what does that sequence mean? A gene turns out to be a recipe for building a

EACH ANIMAL SPECIES HAS A DIFFERENT GENOME WITH A DIFFERENT NUMBER OF CHROMOSOMES AND A DIFFERENT ARRANGEMENT OF GENES ON THOSE CHROMOSOMES. THE DIFFERENCES BETWEEN GENOMES LEADS TO VERY DIFFERENT BODY STRUCTURES AND DEVELOPMENTAL PROCESSES BETWEEN SPECIES. INDIVIDUALS OF THE SAME SPECIES HAVE VERY SIMILAR GENOMES CARRYING THE SAME GENES, THOUGH GENES CAN HAVE SLIGHTLY DIFFERENT FORMS, CALLED ALLELES, THAT CAUSE DIFFERENCES IN PARTICULAR TRAITS.

particular protein. Proteins are large molecules made up of smaller units called amino acids in particular sequences. All life on Earth is made up of twenty amino acids, although many more are possible.

When the organism needs a particular protein, a particular gene contains the recipe to build it as a series of instructions. Each instruction is a series of three "letters," which means there can be a total of sixty-four different ones. Because of the way bases pair, there are actually only thirty-two different possibilities. ATC on one rail of the ladder has a matching TAG on the other, so both combinations mean the same thing to the protein-building cells.

Most instructions say, "Add this amino acid to the protein you are building." But the building has to end at some point, so at least one instruction must mean, "Stop." The thirty-two combinations are more than enough for twenty amino acid choices and one stop command, so there are some duplicates.

In many ways, a gene is like a computer program, and cells are like computers that operate on chemical instructions instead of electrical signals. Although Charles Darwin never imagined that DNA lay at the heart of evolution, he would understand it all if he were alive today. All the genomes of all the species that have ever existed evolved step by step from the very first, very simple living organisms.

That seems hard to believe—until you realize that this evolution has taken place over billions of years in trillions of organisms and billions of species. Each individual has been just a little different from its ancestors. The sum of all those small changes is life on Earth as we know it today.

CHAPTER SIX

AFTER THE BREAKTHROUGH

It took many years of research to get from Watson and Crick's description of the DNA double helix to being able to decode its instructions. The first steps were for other researchers, including Wilkins and Pauling, to make further tests to confirm the structure of DNA that Watson and Crick had proposed. Although some minor corrections were needed, their fundamental work remained correct.

RECOGNITION

Once Crick and Watson's results withstood the scrutiny of others, there was little doubt that their work was worthy of the Nobel Prize in Physiology or Medicine. In 1962, Francis Crick, James Watson, and Maurice Wilkins shared that award "for their discoveries concerning the molecular structure of

FRANKLIN'S CANCER

Rosalind Franklin was remarkably careful about her research but remarkably careless in one respect. By the 1950s, it was well known that high exposure to X-rays can cause cancer. Yet Franklin would do whatever it took to get an excellent X-ray image, even if it meant exposing herself as well.

The X-ray exposure may well have cost her life. In the summer of 1956, while traveling and speaking in the United States, she suffered a serious attack of abdominal pain. She went to a doctor for painkillers and continued her trip. There was too much to do to go to the hospital. She even extended her visit, and by the time she saw a doctor in England, it was too late. She had advanced ovarian cancer.

Through treatment and surgery, she kept working until only days before her death on April 16, 1958.

nucleic acids and its significance for information transfer in living material."

Why Wilkins and not Franklin? The rules for awarding the Nobel Prize state that it is to go only to living people, and Franklin died tragically of cancer in 1958 when she was only thirty-seven.

Had Franklin lived, the Nobel committee might have faced a difficult choice. The rules say that Nobel Prizes for science may never be divided more than three ways. Was she more deserving than Wilkins? Would the committee have elected to give it to Crick and Watson only?

These questions will never be answered, although the Nobel Prize acceptance speeches that year might provide a hint. Neither Crick nor Watson mentioned Franklin's name. Only Wilkins acknowledged "my late colleague Rosalind Franklin who, with great ability and experience of X-ray diffraction, so much helped the initial investigations on DNA."

SEPARATE PATHS

Wilkins remained at King's College for his entire career. He retired in 1981 and died on October 5, 2004. Watson and Crick worked together briefly but soon took separate paths.

Crick focused on the relationship between DNA and proteins, an area in which he made significant discoveries over many years. In 1976, he went to work

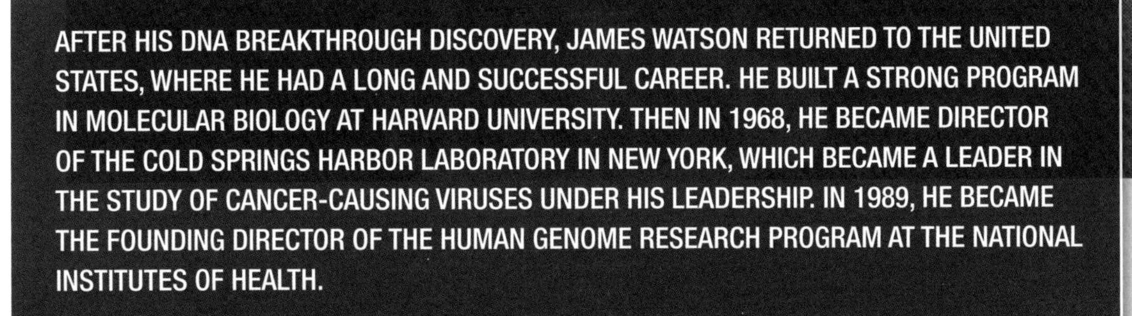

AFTER HIS DNA BREAKTHROUGH DISCOVERY, JAMES WATSON RETURNED TO THE UNITED STATES, WHERE HE HAD A LONG AND SUCCESSFUL CAREER. HE BUILT A STRONG PROGRAM IN MOLECULAR BIOLOGY AT HARVARD UNIVERSITY. THEN IN 1968, HE BECAME DIRECTOR OF THE COLD SPRINGS HARBOR LABORATORY IN NEW YORK, WHICH BECAME A LEADER IN THE STUDY OF CANCER-CAUSING VIRUSES UNDER HIS LEADERSHIP. IN 1989, HE BECAME THE FOUNDING DIRECTOR OF THE HUMAN GENOME RESEARCH PROGRAM AT THE NATIONAL INSTITUTES OF HEALTH.

at the Salk Institute for Biological Studies in San Diego, California, where he studied the brain and pursued the nature of consciousness for the rest of his life. He died on July 28, 2004.

Watson helped build a strong school of molecular biology at Harvard University, Cambridge, Massachusetts. In 1968, he became the director of the Cold Springs Harbor Laboratory, New York, where he raised large amounts of money to save the lab and steered it into the field of tumor virology (the study of cancer-causing viruses). In 1994, he was made the president of CSHL.

He was instrumental in starting the Human Genome Project at the U.S. National Institutes of Health and was director of the project from 1989 to 1992. Watson also created controversy through his writing. Crick and Wilkins objected to his account of their work in *The Double Helix: A Personal Account of the Discovery and Structure of DNA*, and many others viewed the book's treatment of Franklin as sexist. He presented her not as an actual person but as the fictionalized "Rosy" with a very difficult personality.

As a result of those objections, Harvard University Press, which had planned to publish the book, dropped it. It was picked up by two other publishers, one in the United States and one in England, and became a best seller.

CHAPTER SEVEN

THE REVOLUTION TO COME

The discovery of the double helix and the four-letter genetic code began a revolution in biology. Scientists developed techniques to clip out portions of that code and to read an organism's genes. Because certain human diseases and conditions were known to be genetic, the ability to identify and read the code of abnormal genes may lead to treatment or cures.

THE HUMAN GENOME PROJECT

The greatest medical advances in the twenty-first century are likely to be due to our knowledge of genes. Some of the credit for that goes to James Watson for his vision of creating the Human Genome Project (HGP), for which he was the first director.

WOULD YOU HAVE YOUR GENOME SEQUENCED?

The cost of sequencing an individual's genome has come down dramatically. In 2013, numerous articles stated that the age of the $1,000 genome was almost at hand. For about a week's income, many middle-class individuals could afford to get their complete personal genome. That could be very valuable for medical treatment or changes in lifestyle. But it could also reveal if it is likely for the person to develop untreatable conditions. For some people, knowing that they have a higher-than-average chance of developing a serious condition could lead them to poor or risky life choices.

Would you pay $1,000 to know your complete genome? That is a difficult question that you may face as an adult.

That project mapped the DNA of several people letter by letter. The result was the complete sequence of human DNA—the genetic code for every gene that makes us human, including different alleles for some genes. It was one of the largest research projects ever conceived. Led by the U.S. National Institutes of Health but with support from international health and medical agencies, the project was officially launched in 1990, with a goal of completing the first complete map of the human genome within fifteen years. Watson was hired in 1989 to serve as its director and oversaw it until 1992.

As the HGP did its work, the technology for sequencing genes became much less expensive. Entrepreneurs, most notably Craig Venter (1946–), saw opportunities to create valuable scientific equipment and gene-based products. He founded Celera Genomics, which sped up the process of mapping genes. On June 26, 2000, with Venter and HGP director Francis Collins (1950–) at his side, U.S. president Bill Clinton (1946–) announced that the first draft of the human genome had been completed. On April 14, 2003, the HGP held a news conference to announce the full genome and the official completion of its mission.

PROMISE AND CONTROVERSY

The end of the Human Genome Project does not mark the end of DNA research. Instead it marks the beginning of a revolution in medicine. In the treatment of cancer, for instance, chemotherapy is a very difficult but often

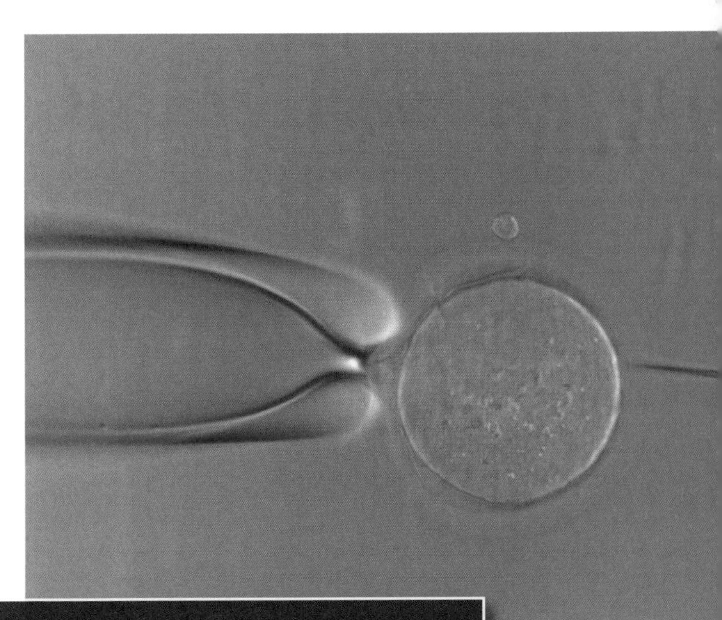

KNOWLEDGE OF THE GENETIC CODE HAS LED TO A REVOLUTION IN MEDICINE, INCLUDING CONTROVERSIAL NEW TECHNIQUES FOR THE TREATMENT OF GENETIC DISEASES. THIS PICTURE SHOWS A RESEARCHER INJECTING DNA INTO A MOUSE EGG.

lifesaving course of treatment. Many patients respond to it and are cured, but others show no improvement. Often the difference in response is genetic. Testing patients' DNA can lead to better choices of treatment.

In some conditions, it may be possible to insert new genes into a person's cells to replace malfunctioning

WHEN WATSON, CRICK, AND WILKINS WERE AWARDED THE 1962 NOBEL PRIZE FOR MEDICINE OR PHYSIOLOGY IN STOCKHOLM, SWEDEN, THEY SHARED THE OCCASION WITH FAMED AUTHOR JOHN STEINBECK, WHO WON THE LITERATURE PRIZE, AND CHEMISTRY WINNERS MAX PERUTZ AND JOHN C. KENDREW. WATSON HAD BEEN ASSIGNED TO WORK WITH PERUTZ WHEN HE JOINED THE CAVENDISH LABORATORY IN 1951. THIS PICTURE SHOWS, FROM LEFT TO RIGHT, WILKINS, PERUTZ, CRICK, STEINBECK, WATSON, AND KENDREW.

ones or tailor a treatment to match a patient's DNA. It is also possible to insert human genes into bacteria or animals so that they can produce useful substances such as insulin to treat human diseases.

The process of inserting genes from one organism into another is called genetic engineering or genetic modification ("gene splicing" for short). It is already finding its way into industrial or agricultural applications. It is often controversial. Some people object to it on religious grounds. Others are concerned that there may be unexpected harmful consequences.

Many people ask if we should move ahead with genetic engineering, but that is the wrong question. Science and business are moving ahead because the promises are great. It is up to the rest of us to ask questions and for societies to develop approaches that weigh the promises against the risks. Those approaches will need to consider the goals and expectations, as well as the religious, moral, and ethical views of our fellow citizens. Finding the best path to our future has never been easy, and that is particularly true with the technological possibilities opened by DNA. Making wise choices about how to use our knowledge of the double helix will be up to the citizens of the future.

TIMELINE

1859 Charles Darwin publishes *On the Origin of Species*, describing evolution by natural selection.

1866 Gregor Mendel publishes the results of his study of heredity in peas.

1869 Friedrich Miescher discovers nuclein, which we now know is DNA, in the nucleus of living cells.

1882 Walther Flemming discovers chromosomes while studying cell division.

1900 Mendel's work is rediscovered.

1907 Thomas Hunt Morgan begins his studies of fruit flies that show that genes are carried on chromosomes.

1916 Francis Crick is born on June 8 in Northampton, England; Maurice Wilkins is born on December 15 in Pongaroa, New Zealand.

1920 Rosalind Franklin is born on July 25 in London, England.

1928 James Watson is born in Chicago, Illinois, on April 6.

1937 Crick graduates from University College, London.

1938 Wilkins graduates from St. John's College of Cambridge University; Franklin enrolls in Newnham College of Cambridge University.

1940 Wilkins completes Ph.D. research at Cambridge.

1943 Watson enters the University of Chicago.

1944 Oswald Avery, Maclyn McCarty, and Colin MacLeod show that DNA carries the genetic code.

1946 Wilkins moves to King's College.

1947 Crick goes to Cambridge to work at Strangeways Laboratory; after completing her Ph.D. at Cambridge, Franklin joins a French government laboratory to study coal.

1949 Crick transfers to Cavendish Laboratory, Cambridge.

1950 Watson graduates from the University of Indiana with a Ph.D. in Zoology.

1951

January 5 - Franklin joins King's College.

May 22-25 - Watson discovers X-ray crystallography in a presentation by Wilkins.

Mid-October - Watson joins Cavendish Laboratory and meets Crick.

November 21 - Watson hears Wilkins and Franklin speak about their work on X-ray crystallography of DNA at a conference in London.

November 26 - Watson and Crick build a triple-helix model of DNA.

November 27 - Franklin sees the DNA model and severely criticizes its errors.

December - Sir Lawrence Bragg, supervisor of the Cavendish, orders Watson and Crick to stop working on DNA.

1952

May 2 - Franklin takes the now famous Photo 51, a remarkably clear and revealing X-ray diffraction image of the B (wet) form of DNA.

Late September - Peter Pauling and Jerry Donohue arrive at the Cavendish.

December 17 (approx.) - Peter Pauling receives a letter from his father, Linus Pauling, that the elder Pauling has found the structure of DNA.

1953

January 28 - Peter Pauling receives advance manuscript copies of his father's paper on the structure of DNA. Watson reads this paper and realizes Pauling has made a major mistake.

January 30 - Watson visits King's College, London, and Wilkins shows him Photo 51. Watson immediately recognizes that is a helical structure, probably a double helix.

January 31 - Bragg gives Watson permission to start building DNA models again.

February 19 - Jerry Donohue corrects Watson on the shapes of the DNA bases.

February 28 - While working with his cardboard cutouts of the four DNA bases, Watson hits upon the correct pairings.

March 7 - Watson and Crick finish assembling their large-scale model.

April 1 - Bragg authorizes Watson and Crick to publish their discovery.

April 25 - Watson's and Crick's paper, "A Structure for Deoxyribose Nucleic Acid," is published in the scientific journal *Nature*. Wilkins, Franklin, and their colleagues publish their X-ray crystallography results in two papers in the same issue.

May 30 - Watson and Crick publish a second article in *Nature*, making further observations on how DNA works.

1958 Franklin dies of ovarian cancer, possibly caused by excessive exposure to X-rays.

1962 Watson, Crick, and Wilkins share the Nobel Prize for Physiology or Medicine.

1968 Watson's book, *The Double Helix*, is published. Watson becomes director of Cold Spring Harbor Laboratory, New York.

1977 Crick begins doing research in the brain and consciousness at the Salk Institute.

1981 Wilkins retires from King's College

1988 Crick's book, *What Mad Pursuit*, is published.

1989 Watson becomes the director of the Human Genome Project, which officially begins in 1990.

1992 Watson resigns from the Human Genome Project. He becomes the president of Cold Spring Harbor Lab.

2000 First draft of human genome announced.

2003 Human Genome Project ends with publication of the full genome.

2004 Crick dies on July 28; Wilkins dies on October 5.

2013 Price for sequencing an individual human genome drops to about $1,000, making it affordable for most middle-class people.

GLOSSARY

ADENINE One of the four chemical bases that make up the structure of DNA.

ARCHAEA (PL.) A kingdom of the most ancient life-forms that, like bacteria, do not contain nuclei or other organelles.

BACTERIUM (PL. BACTERIA) A microscopic, single-celled organism that does not contain a nucleus or other organelles.

BIOLOGY The study of living organisms.

CHROMOSOME A thread-like structure made up of DNA and protein that appears when cells divide.

CYTOSINE One of the four chemical bases that make up the structure of DNA.

DNA Deoxyribonucleic acid; a long, thin molecule found in the nucleus of a cell.

EVOLUTION The natural process through which species change and develop over time.

FUNGUS (PL. FUNGI) A kingdom of life that includes yeast and mushroom.

GENE A section of DNA within a chromosome, which contains genetic code that causes a certain type of protein to be made.

GENETIC Having to do with genes or heredity.

GENETICS The study of heredity and the variation of inherited characteristics.

GENOME All of the genetic material in the nucleus of a cell.

GUANINE One of the four chemical bases that make up the structure of DNA.

HEREDITY The passing on of physical or other traits through the genetic code from one generation to another.

MOLECULE The smallest fundamental unit of a chemical compound.

MOLECULAR BIOLOGY The study of molecules within living organisms.

NATURAL SELECTION The mechanism that leads to evolution, in which certain individuals that are better adapted to their environment are more likely to reproduce their traits in the next generation.

NUCLEOID REGION The central region of bacteria and archaea that contains the organism's DNA.

NUCLEOTIDE A three-part chemical unit made up of a sugar molecule, a phosphate molecule, and (in DNA) one of the four bases: adenine, guanine, thymine or cytosine.

NUCLEUS (PL. NUCLEI) In biology, the spherical sac in the middle of a cell that contains DNA; in physics, the tiny central region of an atom.

ORGANIC CHEMISTRY The study of matter (specifically containing carbon atoms) mostly derived from natural sources and the chemical changes this matter undergoes.

ORGANISM Any living thing.

PHOSPHATE One phosphorous atom and four oxygen atoms bound together to act as one unit.

PHYSICS The study of the properties and interactions of matter and energy.

PROTEIN Long-chained molecules used to build and repair cells.

PROTISTS A family of organisms, mainly one-celled, that contain nuclei and other organelles, which are different from animals, plants, and fungi in that their cells do not organize into structures.

SUGAR A class of sweet-flavored chemicals composed of carbon, hydrogen, and oxygen, usually with two hydrogen atoms for each oxygen. The special sugar that is found in DNA consists of five carbon atoms, six hydrogen, and three oxygen atoms.

THYMINE One of the four chemical bases that make up the structure of DNA.

TRAIT A physical or behavioral characteristic of an organism.

TRANSISTOR A solid-state electronic device that is used to control the flow of electricity in electronic equipment.

X-RAY CRYSTALLOGRAPHY A technique that uses X-ray diffraction to reveal the arrangement of atoms or molecules inside crystals.

X-RAY DIFFRACTION A process in which X-rays reflect off layers in a crystal to produce a photographic image.

ZOOLOGY The subfield of bilogy that deals with the animal kingdom.

FOR MORE INFORMATION

Cold Spring Harbor Laboratory
P.O. Box 100
1 Bungtown Road
Cold Spring Harbor, NY 11724
(516) 367-8455
Web site: http://www.cshl.edu/campus-events/campus-tours/tour-cshl.html

This laboratory, which was led for many years by James Watson, offers public tours and events on an ongoing basis.

Life Sciences-HHMI Outreach Program
Department of Molecular and Cellular Biology
Harvard University
BioLabs 1090 16 Divinity Avenue
Cambridge, MA 02138
(617) 496-3457
Web site: http://outreach.mcb.harvard.edu

James Watson spent many years at Harvard University. This program provides outreach to high schools with educational materials and on-site programs for teachers and students.

Museum of Science and Industry
57th Street and Lake Shore Drive
Chicago, IL 60637-2093
(773) 684-1414
Web site: http://www.msichicago.org

One of the leading science museums in the United States, the Museum of Science and Industry has a large set of biology exhibits, including some on genetics and DNA.

National Museum of Natural History
Smithsonian Institution
Constitution Avenue NW
Washington DC 20560
(202) 633-1000
Web site: http://www.mnh.si.edu
The National Museum of Natural History is part of the Smithsonian Institution. It has a variety of permanent exhibits on the life sciences.

Ontario Science Center
770 Don Mills Road
Toronto, ON M3C 1T3
Canada
(416) 696-1000
Web Site: http://www.ontariosciencecentre.ca
The Ontario Science Centre is one of North America's most popular and innovative science museums, with numerous exhibits and programs on the life sciences and technology.

Salk Institute for Biological Studies
10010 North Torrey Pines Road
La Jolla, CA 92037

(858) 453-4100
Web site: http://www.salk.edu
The Salk Institute, where Crick moved in 1977 to study the brain and consciousness, has an educational outreach program that aims to bring the excitement of science and scientific discovery to the San Diego community and to middle schools and high schools in the area.

Tech Museum of Innovation
201 South Market Street
San Jose, CA 95113
(408) 294-8324
Web site http://www.thetech.org
This museum focuses on technology and innovation, including the latest developments in genetics. Its mission is "to inspire the innovator in everyone."

WEB SITES

Due to the changing nature of Internet links, the Rosen Publishing Group, Inc., has developed an online list of Web sites related to the subject of this book. This site is updated regularly. Please use this link to access the list:

http://www.rosenlinks.com/RDSP/wats

FOR FURTHER READING

BOOKS BY CRICK, WATSON, AND WILKINS

Crick, Francis. *Astonishing Hypothesis: The Scientific Search for the Soul*. New York, NY: Scribner, 1995.

Crick, Francis. *Life Itself*. New York, NY: Simon & Schuster, 1981.

Watson, James D. *Genes, Girls, and Gamow: After the Double Helix*. New York, NY: Knopf, 2002.

Watson, James D. *A Passion for DNA: Genes, Genomes, and Society*. Cold Spring Harbor, NY: Cold Spring Harbor Laboratory Press, 2001.

Watson, James D., and Andrew Berry. *DNA : The Secret of Life*. New York, NY: Knopf, 2003.

Wilkins, Maurice. *The Third Man of the Double Helix*. Oxford, England: Oxford University Press, 2003.

BOOKS BY OTHER AUTHORS

Anderson, Lara. *Rosalind Franklin*. Chicago, IL: Raintree, 2009.

Anderson, Michael. *A Closer Look at Genes and Genetic Engineering*. New York, NY: Britannica Educational Publishing, 2012.

Berlatsky, Noah. *Genetic Engineering*. Detroit, MI: Greenhaven Press, 2013.

Bortz, Fred. "Mapping What Makes Us Human: The Human Genome Project." *To the Young Scientist: Reflections on Doing and Living Science*. New York, NY: Franklin Watts, 1997.

Day, Trevor. *Genetics: Investigating the Function of Genes and the Science of Heredity*. New York, NY: Rosen Publishing, 2013.

DiDomenico, Kelly. *Women Scientists Who Changed the World*. New York, NY: Rosen Publishing, 2012.

Foy, Debbie. *Medical Pioneers*. New York, NY: PowerKids Press, 2011.

Gingerich, Owen. *Rage or Reason? When Scientists Feud*. Peterborough, NH: Cobblestone Publishing, 2011.

Graham, Ian. *Genetics*. Irvine, CA: Saddleback Educational Publishing, 2010.

Green, Jen. *Inheritance and Reproduction*. Chicago, IL: Capstone Heinemann Library, 2014.

Guttman, Burton S. *Genetics: The Code of Life*. New York, NY: Rosen Publishing, 2011.

Hand, Carol. *Introduction to Genetics*. New York, NY: Rosen Publishing, 2011.

Heos, Bridget. *The Human Genome*. New York, NY: Rosen Publishing, 2011.

Hollar, Sherman. *Pioneers in Medicine: From the Classical World to Today*. New York, NY: Britannica Educational Publishing, 2013.

Hyde, Natalie. *DNA*. New York, NY: Crabtree Publishing, 2010.

Lew, Kristi. *Heredity*. New York, NY: Chelsea House, 2009.

Polcovar, Jane. *Rosalind Franklin and the Structure of Life*. Greensboro, NC: Morgan Reynolds, 2006.

Rooney, Anne. *Watson and Crick and Their Discovery of DNA*. London, England: Evans, 2010.

Schultz, Mark, Zander Cannon, and Kevin Cannon. *The Stuff of Life: A Graphic Guide to Genetics and DNA*. New York, NY: Hill and Wang, 2009.

Silverstein, Alvin, Virginia B. Silverstein, and Laura Silverstein Nunn. *DNA*. Minneapolis, MN: Twenty-First Century Books, 2009.

Yount, Lisa. *Rosalind Franklin: Photographing Biomolecules*. New York, NY: Chelsea House, 2011.

BIBLIOGRAPHY

Clarke, Tom. "DNA's Family Tree". Nature.com, 2003. Retrieved April 21, 2013 (http://www.nature.com/nsu/030421/030421-5.html).

Crick, Francis. *What Mad Pursuit*. New York, NY: Basic Books, 1988.

Ehrlich, Paul R. *Human Natures: Genes, Cultures, and the Human Prospect*. Washington, DC: Island Press, 2000.

Genome News Network. "Genetic and Genomics Timeline." 2002–2004. Retrieved April 21, 2013 (http://www.genomenewsnetwork.org/timeline/timeline_home.shtml).

Kean, Sam. *The Violinist's Thumb and Other Lost Tales of Love, War, and Genius as Written by Our Genetic Code*. New York, NY: Little Brown, 2012.

Lasker Medical Research Network. "James Watson Timeline." Retrieved April 21, 2013. (http://www.laskerfoundation.org/awards/kwood/watson/timeline.shtml).

McCarty, Maclyn. "Discovering Genes Are Made of DNA." *Nature* 421, 406 (January 23, 2003).

National Centre for Biotechnology Education. "Double Helix: 1953–2003." 2004. Retrieved April 21, 2013. (http://www.ncbe.reading.ac.uk/DNA50/timeline.html).

Public Broadcasting System. "The Thousand Dollar Genome." January 25, 2013. Retrieved April 9, 2013 (http://www.pbs.org/wnet/religionandethics/episodes/january-25-2013/the-thousand-dollar-genome/14569).

Ridley, Matt. *Genome: The Autobiography of a Species in 23 Chapters*. New York, NY: HarperCollins, 2000.

Stock, Gregory. *Redesigning Humans: Our Inevitable Genetic Future*. New York, NY: Houghton Mifflin, 2002.

Tudge, Colin. *The Impact of the Gene: From Mendel's Peas to Designer Babies*. New York, NY: Hill and Wang, 2001.

U.S. Department of Energy Office of Science. "Human Genome Project Information." Retrieved April 9, 2013 (http://www.ornl.gov/sci/techresources/Human_Genome/home.shtml)

Wade, Nicholas. "DNA, the Keeper of Life's Secrets, Starts to Talk." *New York Times*, February 25, 2003. Retrieved April 21, 2013 (http://www.nytimes.com/2003/02/25/science/25HELI.html?ex=1082692800&en=b79c030ff28e3f25&ei=5070)

Watson, James D. *The Double Helix*. New York, NY: Penguin Books, 1968.

Watson, James D. "Nobel Prize Banquet Speech," 1962. Retrieved April 29, 2013 (http://www.nobel.se/medicine/laureates/1962/watson-speech.html).

INDEX

K

M

N

P

R

V

W

X

ABOUT THE AUTHOR

R. N. Albright is the author of numerous science books for young readers.

PHOTO CREDITS

Cover (portraits) Source-Science/Photo Researchers/ Getty Images; cover (DNA strand) © iStockphoto.com/ geopaul; p. 5 Comstock/Thinkstock; p. 7 Hulton Archive/Getty Images; pp. 10, 55 Science Source/ Photo Researchers/Getty Images; p. 12 Express/Archive Photos/Getty Images; p. 15 George Silk/Time & Life Pictures/Getty Images; p. 20 © AP Images; p. 24 The Bridgeman Art Library/Getty Images; pp. 26, 34 A. Barrington Brown/Science Source; p. 28 Tom Hollyman/ Photo Researchers/Getty Images; p. 31 Leonard Lessin/ Photo Researchers/Getty Images; p. 40 Science Source; p. 45 Science & Society Picture Library/Getty Images; p. 48 J. L. Carson/Custom Medical Stock Photo/Getty Images; p. 52 Anna Jurkovska/Shutterstock.com; p. 57 Andreas Feininger/Time & Life Pictures/Getty Images; p. 61 Michel Baret/Gamma-Rapho/Getty Images; p. 62 Keystone/Hulton Archive/Getty Images; cover and interior pages (textured background) © iStockphoto.com/Perry Kroll, (atom illustrations) © iStockphoto.com/suprun.

Designer: Nicole Russo; Photo Researcher: Karen Huang